The Ultimate Guide to Tilapia Farming

A Step By Step Guide to Raise Your Fish till Maturity

A Comprehensive Guide to Caring, Breeding, Nutrition, Health, Behavioral Understanding, and Lifelong Well-Being for Your Tilapia Fish

Dr. Spencer Sean

DISCLIAMER

The information provided in this book, "The Ultimate Guide to Tilapia Farming," is intended for general informational purposes only. It is based on the author's research, experience, and knowledge at the time of writing. While every effort has been made to ensure the accuracy and completeness of the content, the author and publisher make no guarantees regarding the results you may achieve from applying any of the information contained herein.

Readers are encouraged to consult with professional advisors, including but not limited to aquaculture specialists, veterinarians, and legal experts, before implementing any of the practices or recommendations described in this book. The techniques and methods discussed may not be suitable for every individual or situation, and the author and publisher disclaim any liability for any adverse effects that may result from the use or application of the information provided.

The author does not endorse, recommend, or have any affiliation with any specific individual, product, website, organization, or other names that may be referenced or mentioned in this book. Any references to such entities are provided solely for informational purposes and should not be construed as endorsements.

The reader assumes full responsibility for any actions taken based on the information presented in this book. The author and publisher shall not be held liable for any loss or damage, including but not limited to direct, indirect, incidental, consequential, or punitive damages, arising out of or related to the use of or reliance on this book.

Copyright © 2024 by Spencer Sean

All rights reserved. No part of this book may be reproduced, distributed, or transmitted in any form or by any means, including photocopying, recording, or other electronic or mechanical methods, without the prior written permission of the publisher, except in the case of brief quotations embodied in critical reviews and certain other non-commercial uses permitted by copyright law.

Published by Spencer Sean

First Edition: 2024

Printed in the United States of America

ABOUT THIS BOOK

The Ultimate Guide to Tilapia Farming is an indispensable resource for anyone looking to delve into the lucrative and sustainable practice of tilapia aquaculture. This comprehensive guide covers every facet of tilapia farming, from its historical roots to the latest innovations and future trends, providing both novice and experienced farmers with the knowledge needed to succeed.

Tilapia farming has long been recognized for its importance due to its role in food security and economic development. This guide begins by delving into the history of tilapia farming, illustrating how this practice has evolved over the years and its significance in today's aquaculture industry. Understanding why rearing tilapia is beneficial sets the stage for aspiring farmers, highlighting the nutritional and economic advantages of cultivating this resilient and fast-growing species.

One of the fundamental aspects covered is the biology of tilapia. Knowing the biological characteristics of different tilapia species is crucial for selecting the right type to rear. The guide provides detailed insights into setting up the perfect habitat, emphasizing water quality management and the necessary physical, chemical, and biological parameters. This ensures that tilapia are raised in an environment conducive to their health and growth, which is further supported by guidance on choosing the best equipment for tilapia rearing.

Feeding and nutrition are pivotal to successful tilapia farming. The book elaborates on basic feeding practices and advanced nutrition strategies to optimize growth, ensuring that tilapia receive a balanced diet that promotes health and productivity. Alongside nutrition, understanding tilapia behavior and managing their health are critical. The guide covers common diseases and their treatments, helping farmers maintain a healthy stock.

The guide also addresses the critical issue of stress management in tilapia, as stress can significantly impact growth and health. It discusses water filtration, aeration systems, and pond management techniques, offering a comparative analysis of indoor versus outdoor rearing to help farmers choose the best method for their circumstances.

A detailed explanation of the ten growth stages of tilapia provides a roadmap for farmers to follow, ensuring they know what to expect at each stage. This is complemented by practical advice on harvesting, processing, and storing tilapia, ensuring that farmers can maximize the value of their yield.

Sustainable farming practices are a cornerstone of modern aquaculture, and this guide dedicates a section to these practices, along with economic considerations for tilapia farming. Understanding the legal and regulatory implications is also crucial, as compliance with local and international standards can impact the viability of a farming operation.

Innovations in tilapia farming, integrating tilapia into other aquaculture systems, and future trends are explored, offering a glimpse into the evolving landscape of tilapia aquaculture. The community and social advantages of tilapia rearing underscore its broader impact beyond individual farming operations.

Finally, the guide provides a comprehensive FAQ section, addressing common questions and concerns that tilapia farmers may have, ensuring that readers have a thorough understanding of all aspects of tilapia farming. This book is not just a manual but a complete reference for sustainable and profitable tilapia aquaculture.

Table of Contents

DISCLIAMER 2
ABOUT THIS BOOK 4
Introduction 11
The Importance of Rearing Tilapia.........................11
History of Tilapia Farming 13
CHAPTER ONE 14
Why Should You Rear Tilapia?................................14
Things to Consider Before Beginning Tilapia Rearing
....15
Understanding Tilapia Biology 17
Ways to Select the Right Tilapia Species 17
CHAPTER TWO 19
Setting up Your Tilapia Habitat 19
Water Quality & Management...............................22
Physical parameters:..22
Chemical Parameters: 22
Biological parameters: 23
Choosing the Best Equipment for Tilapia Rearing .24
CHAPTER THREE 27

Understanding Tilapia Behavior 27

Feeding and Nutrition Basics 28

Advanced nutrition for optimal growth 30

CHAPTER FOUR 33

Managing Tilapia Health 33

17 Common Tilapia Diseases and Treatments 35

CHAPTER FIVE 40

Understanding and managing stress in tilapia 40

CHAPTER SIX 45

Water filtration and aeration systems 45

Pond Management for Tilapia 47

Indoor vs. Outdoor Tilapia Rearing 50

CHAPTER SEVEN 55

Ten Tilapia Growth Stages 55

Steps to Follow When Harvesting Tilapia 56

Ways to Process and Store Tilapia 58

CHAPTER EIGHT 60

Sustainable Tilapia Farming Practices 60

Economic Considerations for Tilapia Farming 62

CHAPTER NINE 68

Legal and Regulatory Implications of Tilapia Farming...68

Innovations in Tilapia Farming 68

Integrating Tilapia into Other Aquaculture Systems 69

CHAPTER TEN 83

Future Trends in Tilapia Farming 83

Community and Social Advantages of Tilapia Rearing 86

CHAPTER ELEVEN 89

Frequently Asked Questions About Tilapia Rearing 89

Conclusion..92

Introduction

Tilapia rearing has become more important in aquaculture because of its economic and nutritional worth. Understanding its history, biology, and practical applications is critical for successful farming.

The Importance of Rearing Tilapia

1. **Nutritional Value:** Tilapia is high in protein, low in fat, and contains critical vitamins and minerals, making it a useful dietary source worldwide.

2. **Economic Impact:** Tilapia cultivation provides livelihoods in many areas, boosting local economies via employment and export prospects.

3. **Environmental Benefits:** When compared to terrestrial animals, tilapia aquaculture has a reduced environmental effect, notably in terms of water consumption and greenhouse gas emissions.

4. **Food Security:** Tilapia production contributes to the growing worldwide demand for fish by providing a sustainable protein source.

5. Diversification enables farmers to diversify their revenue sources and minimize their reliance on conventional agriculture.

6. **Cultural Importance:** Tilapia is culturally significant in various civilizations, contributing to culinary traditions and regional cuisines.

7. Tilapia's omega-3 fatty acid concentration may help enhance cardiovascular health when consumed regularly.

8. **Aquatic Ecosystem Management:** Integrated farming systems based on tilapia may help manage and use natural resources more effectively.

9. Tilapia farming promotes continuous research in aquaculture technology, genetics, and sustainable practices.

10. Global Trade: Tilapia is a prominent participant in the international seafood trade, with significant market impact.

History of Tilapia Farming

Tilapia raising stretches back thousands of years, with historical evidence pointing to its cultivation in ancient Egypt. Tilapia was first cultivated for its ease of breeding and quick growth in warm regions, and it spread widely via trade routes. Modern aquaculture methods have improved tilapia farming efficiency and sustainability.

CHAPTER ONE

Why Should You Rear Tilapia?

1. Tilapia grows fast, reaching market size in 6-9 months, allowing for high turnover and profitability.

2. **Adaptability:** Tilapia thrives in a variety of aquatic habitats and can tolerate a broad range of water quality and temperature.

3. **Low Trophic Level:** Tilapia, being omnivores, may graze on natural algae and organic materials, lowering feed costs and environmental effects.

4. Tilapia has steady demand in both home and foreign markets, providing reliable market prospects.

5. Tilapia farming requires little space and may be conducted in tiny ponds or controlled conditions, making it appropriate for a wide range of agricultural sizes.

6. **Disease Resistance:** Certain tilapia species are very resistant to common aquatic illnesses, which reduces health management issues.

7. **Breeding Potential:** Tilapia are prolific breeders, allowing for self-sustaining populations and reduced reliance on external inputs.

8. **Environmental Compatibility:** Well-managed tilapia aquaculture can coexist with existing ecosystems, promoting biodiversity and natural resource usage.

Things to Consider Before Beginning Tilapia Rearing

1. Regulatory Requirements: Become acquainted with local rules and permits for aquaculture operations.

2. **Market Analysis:** Conduct extensive market research to better understand demand, price patterns, and market channels.

3. Assess your infrastructure needs, such as ponds, tanks, or recirculating aquaculture systems (RAS).

4. **Water Quality Management:** Establish methods for monitoring and managing water quality, as well as ensuring access to clean water sources.

5. **Feed and Nutrition:** Determine the best feed sources and nutrition management for each tilapia species and development stage.

6. **Disease Management:** Implement biosecurity measures and prepare for disease outbreaks.

7. Financial planning includes estimating starting costs, ongoing expenditures, and possible income sources to construct a successful company strategy.

8. **Labor and Skills:** Determine workforce requirements and provide sufficient training for tilapia growing operations.

9. **Environmental Impact:** Consider your farm's ecological imprint and use sustainable methods when possible.

10. **Risk Assessment:** Determine possible risks such as weather, market volatility, and regulatory changes, and devise mitigation solutions.

Understanding Tilapia Biology

Tilapia are freshwater fish from the Cichlidae family, notable for their adaptability and reproductive abilities. They display a wide range of behaviors and eating patterns according to the species and environmental circumstances, making them suited for a variety of agricultural techniques.

Ways to Select the Right Tilapia Species

1. **Area Compatibility:** Select species that flourish in your local area, whether tropical or temperate.

2. **Market Preference:** Choose species that are in great demand and have stable prices in your target market.

3. Consider species with quick growth rates to improve production efficiency and profitability.

4. Size and Yield: Determine the average size at harvest and yield potential per unit of farmed area or volume.

5. Disease Resistance: Choose species that have been verified as resistant to common illnesses in your agricultural location.

6. Water Quality Tolerance: Determine which species can withstand fluctuations in water quality parameters characteristic of your agricultural location.

CHAPTER TWO

Setting up Your Tilapia Habitat

Tilapias thrive in warm water, so choose a site that gets enough of sunshine. Ensure that the region is pollution-free and has simple access to power and water.

- **Choosing the Right Tank or Pond:** Determine whether you'll use a tank or a pond. Ponds are more natural and bigger, while tanks provide more control over circumstances. Ensure that the size is appropriate for the amount of fish.

- **Filling the Habitat:** Fill the tank or pond with clean, dechlorinated water. Use water from a reputable source to ensure it is free of toxins.

- **Installing Aeration Systems:** Tilapia need oxygen to thrive. Install aeration devices to maintain enough oxygen levels in the water. This might include air pumps, diffusers, and waterfalls.

- **Setting Up Filtration Systems**: An effective filtration system eliminates trash and pollutants from water. Choose biological filters to keep your water clean and limit the danger of sickness.

- **Monitoring And Adjusting Ph Levels:** Tilapia grow in water with a pH of 6.5 to 8. Water should be tested regularly and the pH adjusted as needed using acceptable chemicals.

- **Controlling Water Temperature:** Keep the water temperature between 75°F and 86°F. Use heaters or chillers as needed to maintain the water within this temperature range.

- **Installing Lighting:** While tilapia do not need specific lighting, they should have a natural light cycle of 12-14 hours each day to simulate their native environment.

- **Creating Hiding Spots and Structures:** To offer hiding places, use structures such as rocks, pipes, or plants. This reduces stress and promotes natural behaviors.
- **Introducing Helpful Bacteria:** Introduce helpful bacteria into the environment to assist break down trash and maintain a healthy ecology. These may be added using commercial items.
- Install a system to periodically check water quality factors including temperature, pH, ammonia, nitrites, and nitrates.
- **Quarantining New Fish:** Before bringing new fish into the ecosystem, quarantine them in a separate tank for a few weeks to guarantee they are disease-free.
- **Regular Habitat Maintenance:** Establish a cleaning and maintenance schedule for the habitat. This involves removing uneaten food, cleaning filters, and checking the water quality.

- **Stocking the Habitat:** Gradually add tilapia to the habitat. Begin with a low number then raise as the system stabilizes.

Water Quality & Management

Water quality is critical to the health and development of tilapia. It entails maintaining an appropriate balance of physical, chemical, and biological elements.

Physical parameters:

Tilapia are tropical fish that need water temperatures ranging from 75°F to 86°F. Temperature variations may stress fish and make them more prone to sickness.

Turbidity: Clear water is necessary. High turbidity may impair eating and gill function.

Chemical Parameters:

pH: Keep the pH level between 6.5 and 8. Test often and modify with lime or acids.

Ammonia, nitrites, and nitrates are all hazardous to fish. Ensure that these are maintained at zero by biological filtration. Nitrates are less hazardous, but they must still be monitored and regulated using water changes or plants.

Biological parameters:

Beneficial Bacteria: These bacteria convert waste materials into less toxic chemicals. Introduce them using commercial goods or by cycling the tank.

Managing water quality:

- **Regular Testing:** Check the water parameters regularly using test kits.
- **Water Changes:** To eliminate waste and replace minerals, do partial water changes regularly.
- **Filtration:** Ensure that the filters are clean and operating properly.
- **Aeration:** Use aerators to keep oxygen levels stable.

- **Biosecurity:** Keep equipment clean and prevent the entrance of sick fish.

Choosing the Best Equipment for Tilapia Rearing

Tanks/Ponds: Based on your available space and budget, choose either tanks or ponds. Ensure that they are big enough to accommodate the amount of fish.

Water pumps are essential for ensuring water circulation and purity. Select a trustworthy pump with appropriate capacity.

Aerators help to keep oxygen levels stable. Air pumps, diffusers, and waterfalls are among the available options.

- **Filters:** To keep water pure, use both biological and mechanical filters. Canister filters are often used in tanks, while bigger systems may need pond filters.

- Heaters and chillers are used to keep water at the proper temperature. Choose based on your climate and the size of your surroundings.
- **Illumination:** Basic illumination to mimic the natural day/night cycle. Avoid excessive or very bright lighting.
- **Test Kits:** Used to evaluate water quality parameters such as pH, ammonia, nitrites, and nitrates.
- **Feeding Equipment:** Automatic feeders provide constant feeding schedules.
- Cleaning tools include nets, scrapers, and siphons for routine maintenance.
- **Thermometers:** Used to check water temperature.
- **Quarantine tank:** Used to isolate new or ill fish.
- Breeding equipment includes spawning nets and crates.

- **Cover/Netting:** Prevents fish from leaping out and keeps predators away.
- **Backup Power Supply**: Ensures that equipment continues to operate during power interruptions.
- Chemicals used in water treatment include pH adjustment, chlorine removal, and ammonia control.

CHAPTER THREE

Understanding Tilapia Behavior

Understanding tilapia behavior is critical for good management. Tilapia are omnivorous and have vigorous eating habits. They often come to the surface to feed.

Schooling: They like to remain in groups, which minimizes stress and promotes development.

Territoriality: Males may be territorial, particularly during breeding. Allow enough room and hiding areas.

Breeding: Females carry eggs in their mouths (mouthbrood). Recognize breeding behaviors to control reproduction.

Stress Indicators: Rapid breathing, irregular swimming, and hunger loss are all signs of stress. Keep an eye out for these symptoms and make any necessary adjustments to the circumstances.

Social Hierarchy: Dominance hierarchies may emerge. Ensure that all fish have access to food and room to avoid bullying.

Feeding and Nutrition Basics

Provide a balanced meal that includes proteins, lipids, and carbs.

Use high-quality commercial feeds formulated specifically for tilapia.

Feed baby fish three to four times each day, and adults once or twice a day.

Quantity: Avoid overeating. Feed just what they can finish in a few minutes.

Supplements: Use vitamins and minerals as required.

Live Food: Provide live food on occasion to provide variety.

Veggies: Serve lush greens and veggies.

Feeding Schedule: Follow a regular feeding schedule.

Observation: Monitor feeding behavior to alter amount and kind.

Avoiding Waste: Remove uneaten food to help keep the water clean.

Adjust your diet based on your development stage.

Energy needs: Meet the energy needs of active fish.

Protein Sources: Include fish meal, soybean meal, and other protein sources.

Fat sources include important fatty acids.

Hydration: Ensure that the meal does not dehydrate the fish.

Advanced nutrition for optimal growth

High-Protein Diets: High-protein feeds promote rapid development.

Amino Acid Profile: Maintain a balanced amino acid profile.

Specialty Feeds: Choose growth-promoting feeds.

Enzyme supplements promote digestion and nutrition absorption.

Probiotics promote intestinal health and nutrition absorption.

Prebiotics: Improve intestinal flora.

- **Mineral Supplements:** Get enough calcium, phosphorus, and other minerals.
- **Vitamin Enrichment:** Include vitamins A, D, E, and B complex.

- Carotenoids improve color and health.
- **Fatty Acid Balance:** Optimize your omega-3 and omega-6 fatty acids.
- **Breeding Tilapia:** A Complete Guide
- **Selecting Breeding Stock:** Choose healthy, adult fish with suitable characteristics.
- **Setting up Breeding Tanks:** Create a separate tank with appropriate parameters.
- **Creating Nesting Sites:** Provide substrate for nesting.
- **Pairing Fish:** In the breeding tank, place male and female fish together.
- Monitor courting and spawning behavior.
- Females nurture eggs in their tongues.
- **Fry Care:** After they are liberated, place them in separate tanks.
- **Feeding Fry:** Use specialized fry food.

- **Growth Monitoring:** Keep track of your growth and wellness.
- **Water Quality:** Maintain ideal water conditions.
- Health monitoring involves keeping an eye out for infections and parasites.
- **Gradual Introduction:** Gradually introduce fry into the main tank.
- Population management entails controlling breeding to prevent overcrowded conditions.
- **Genetic variety:** Maintain genetic variety by combining stocks.

Maintain careful records of breeding and development.

Plan to sell or increase the population.

CHAPTER FOUR

Managing Tilapia Health

Managing tilapia health entails avoiding infections and maintaining ideal environments.

Regular Health Checks: Inspect fish regularly for symptoms of illness or stress.

Always isolate new fish before bringing them into the main environment.

To avoid illnesses, keep the environment clean, including water and habitat.

Proper Nutrition: Provide a well-balanced diet to promote immunity.

Disease Treatment: Use the right therapies for illnesses. Consult a veterinarian if required.

Stress Reduction: To reduce stress, keep water conditions steady and provide appropriate room.

Parasite control entails regularly checking for and treating parasites.

Maintain records of health conditions and treatments for future reference.

Implement biosecurity measures to avoid disease introduction and spread.

Vaccination: If vaccines for certain illnesses are available, consider getting them.

By following these methods and concepts, you may create a healthy, productive environment for tilapia farming.

17 Common Tilapia Diseases and Treatments

1. Bacterial hemorrhagic septicemia is caused by a variety of bacteria, including Aeromonas and Pseudomonas. Symptoms include crimson stripes over the body, a bloated belly, and ulcers. Antibiotics such as oxytetracycline are used in treatment, as well as improved water quality to alleviate fish stress.

2. Columnaris Disease is caused by the bacteria Flavobacterium columnare. Symptoms include white patches on the gills, frayed fins, and skin sores. Treatments include potassium permanganate baths and antibiotics like oxytetracycline.

3. Streptococcosis is a bacterial illness caused by Streptococcus iniae. Symptoms include exophthalmos (bulging eyes), hemorrhages, and irregular swimming. Antibiotic treatments and ensuring adequate water quality are critical to management.

4. Mycobacteriosis, caused by Mycobacterium species, is a chronic condition that causes weight loss, organ nodules, and skin ulcers. There is no effective therapy, although increasing water quality and implementing biosecurity measures may help avoid epidemics.

5. Edwardsiellosis is a condition caused by Edwardsiella tarda that is characterized by ulcerative ulcers, abdominal swelling, and internal bleeding. Antibiotics and better cleanliness are used to treat the condition.

6. Vibrosis is caused by Vibrio species, and symptoms include skin ulcers, fin erosion, and hemorrhage. Antibiotics are used as treatment, as well as preserving ideal water conditions.

7. Aeromoniasis is caused by the pathogen Aeromonas hydrophila, which causes ulcers, fin rot, and systemic infections. Antibiotics and improved water quality are used as treatments to alleviate stress.

8. Ichthyophthirius multifiliis (Ich) is a parasite illness that creates white blotches on the skin and gills. To cure the parasite, increase the temperature of the water and use chemical treatments like as formalin or copper sulfate.

9. Trichodina, a protozoan parasite, causes skin and gill irritation, resulting in flashing and increased mucus production. Formalin and salt baths are excellent therapies.

10. Costia (Ichthyobodo) is another protozoan parasite that causes gill and skin irritation, which results in increased mucus production and lethargy. Formalin and salt baths are popular therapies.

11. Gill flukes (Dactylogyrus) are monogenean parasites that adhere to the gills, causing discomfort and respiratory distress. Praziquantel or formalin baths are used to treat the condition.

12. Skin flukes (Gyrodactylus) are similar to gill flukes but cause skin irritation and increased mucus production. Praziquantel and formalin treatments are successful.

13. Epistylis is a protozoan that clings to the skin and gills, resulting in white patches and excessive mucus. Baths with potassium permanganate or formalin are used as treatment.

14. Fish lice (Argulus) are ectoparasites that cling to the skin, causing discomfort and perhaps secondary illnesses. Chemical therapies, such as dimilin or salt baths, are used during therapy.

15. Anchor Worm (Lernaea): These copepod parasites burrow into the skin, producing sores and perhaps secondary illnesses. Manual removal and chemical treatments, such as organophosphates, are utilized.

16. Saprolegniasis is a fungal illness that produces cotton-like growths on the skin and gills. Treatment

includes increasing water quality and using antifungal medicines such as malachite green.

17. Branchiomycosis is a fungal illness that damages the gills, causing respiratory discomfort and increased mortality. Improving water quality and using antifungal treatments are critical.

CHAPTER FIVE

Understanding and managing stress in tilapia

Water quality monitoring is critical for achieving and maintaining ideal water conditions. pH, ammonia, nitrite, and nitrate levels must be maintained under control to minimize stress. Use water testing kits to modify the water's chemical as required.

2. Temperature Control: Tilapias thrive in warm water. Maintain temperatures of 25°C to 30°C. To avoid temperature-related stress, use heaters in colder areas and shade in hotter places.

3. Proper stocking density: Overcrowding may cause competition for resources and increased stress. Maintain proper stocking densities (20-30 kg/m^3) to guarantee enough space and resources.

4. Adequate Nutrition: Offer a balanced meal that covers all nutritional requirements. High-quality commercial feeds intended for tilapia provide critical nutrients while decreasing stress and increasing development.

5. Routine Health Checks: Check fish regularly for symptoms of sickness or damage. Early diagnosis and treatment of health conditions help prevent stress from rising and spreading across the community.

6. Handling should be kept to a minimum since it might induce tension. Use delicate techniques and handle fish as little as possible. When required, use soft netting to reduce air exposure.

7. Providing Shelter: Use structures like as PVC pipes or aquatic plants to provide hiding places. Shelters minimize aggressiveness and create a feeling of security, which lowers stress levels.

8. Maintaining Cleanliness: Clean tanks or ponds regularly to eliminate garbage and leftover food. A clean

atmosphere minimizes the likelihood of illness and stress. Implement appropriate waste management methods.

9. Aeration ensures enough dissolved oxygen levels. Low oxygen levels create stress and may be lethal. Maintain ideal levels by using aerators or oxygen diffusers.

10. Stable Light Conditions: Avoid rapid illumination changes. Gradual transitions match natural settings, reducing stress. Use timers to control light cycles in indoor systems.

11. Water Flow Management: Make sure the water flow is mild and continuous. Sudden fluctuations in flow rate might cause stress in tilapia. Create a steady aquatic environment by using pumps and filters.

12. Avoiding Chemical Contamination: Keep pollutants and chemicals out of the water. Clean and treat only with safe, certified chemicals. Prevent agricultural runoff and other pollution.

13. Quarantine New Fish: For a few weeks, isolate any new additions to the colony. This method helps to avoid disease transmission and lowers stress caused by unexpected environmental changes.

14. Training Staff: Ensure that all workers are properly taught fish handling and care practices. Staff that are educated may recognize and minimize stressors more effectively.

15. Maintain water quality by doing partial water changes regularly. This helps to eliminate contaminants and freshen the surroundings, lowering stress.

16. Implement strong biosecurity standards to avoid disease outbreaks. Use disinfectants, limit access to raised areas, and keep an eye out for illness symptoms.

17. Probiotics: Include probiotics in the meal to encourage healthy gut flora. Probiotics may improve immune function and prevent stress-related illnesses.

18. Proper Breeding Practices: Avoid inbreeding and maintain genetic variety. Healthy breeding methods produce strong, stress-resistant offspring.

19. Predator Control: Use nets or obstacles to keep tilapia safe from predators. The presence of predators may induce persistent stress and negatively impact overall health.

20. Stress Indicators: Learn to identify stress signals like as irregular swimming, color changes, or decreased eating. To preserve a healthy atmosphere, promptly address any difficulties that arise.

CHAPTER SIX

Water filtration and aeration systems

Water filtration and aeration systems are critical for maintaining high water quality and oxygen levels in tilapia farms. Effective water filtration eliminates physical, chemical, and biological impurities, resulting in a healthier environment. There are many different kinds of filtration systems available, each with its own set of functions.

Mechanical filtering removes solid particles from water. This may be accomplished using screens, filters, or sedimentation tanks. Mechanical filtration prevents waste from accumulating, which may decrease water quality and raise stress in tilapia.

Biological filtration is critical for handling nitrogenous waste, which may be harmful at high concentrations. This method uses helpful microorganisms to convert hazardous ammonia and nitrite into less damaging nitrate via the nitrogen cycle. Biofilters, which are often made of substrates such as bio-balls or sponge filters, give a surface for these bacteria to inhabit. Ensuring enough biological filtration capacity is critical for preserving a balanced and healthy aquatic environment.

Chemical filtration uses activated carbon or other adsorbent materials to extract dissolved organic molecules and contaminants. This form of filtration reduces smells and improves water clarity. To be effective, the chemical filtering medium must be replaced regularly.

Aeration systems are essential for maintaining enough dissolved oxygen levels, which are important for tilapia survival. Low oxygen levels may lead to substantial stress and even death. Aeration may be accomplished using a variety of means, including air stones, diffusers, and motorized aerators.

Air stones and diffusers create fine bubbles in the water, increasing the available surface area for gas exchange. This improves oxygen absorption and facilitates the elimination of CO_2. The placement of air stones or diffusers should guarantee that oxygen is evenly distributed throughout the tank or pond.

Mechanical aerators, such as paddlewheels or venturi systems, provide water movement and turbulence, which improves oxygen transport. These devices are especially successful in bigger ponds and tanks. The aeration technique used is determined by the tilapia farming setup's unique requirements, such as tank size, stocking density, and water flow rates.

Pond Management for Tilapia

Pond maintenance is critical to tilapia farming success. Proper pond management maintains ideal water quality, reduces stress, and encourages healthy development. To keep a tilapia pond productive, many important factors must be addressed.

First, location selection is crucial. Choose a place with a reliable water supply, sufficient soil for pond building, and enough sunshine. The property should also be safeguarded from floods and pollution from agricultural runoff or industrial waste.

Pond construction should address the pond's design, depth, and size. Ideal ponds are rectangular or square and have a depth of 1-2 meters. Adequate slope for drainage and water circulation is required. Ponds should be built for simple harvesting and upkeep.

Water quality control is critical to pond health. Regularly monitor and maintain water characteristics such as pH, temperature, dissolved oxygen, and nutrient concentrations. Liming the pond bottom before filling it with water might assist in maintaining pH levels. Fertilization with organic or inorganic fertilizers may increase primary production and provide natural food sources for tilapia.

Proper stocking density is critical for avoiding overpopulation and providing enough nutrients for each fish. Stocking rates generally vary between 2 and 4 fish per square meter, depending on the production method and management procedures.

Feeding strategies have a substantial influence on pond management. Provide a balanced diet with enough protein levels. Feeding should be done at regular intervals to ensure that the tilapia have enough food without overfeeding, which may lead to water quality concerns.

Regular pond management is required to eliminate excess organic waste, regulate aquatic vegetation, and avoid the accumulation of dangerous compounds. This involves regular fish harvesting to maintain ideal stocking levels and for long-term development.

Disease management entails doing frequent health exams and monitoring for symptoms of sickness. Before adding fresh fish to the pond, quarantine them and treat them as needed if illness is identified. Implement biosecurity measures to prevent infections from spreading.

Pond aeration is critical for maintaining proper oxygen levels, particularly during times of high temperature or stocking density. Mechanical aerators or air diffusers may help to increase oxygenation and avoid hypoxia.

Indoor vs. Outdoor Tilapia Rearing

Tilapia may be grown inside or outdoors, with each having its own set of benefits and problems. Understanding the contrasts between these systems may help you make educated choices about tilapia production.

Indoor Tilapia Rearing

Indoor systems provide controlled settings that enable exact manipulation of water quality, temperature, and light. This management mitigates the danger of environmental swings and predators, resulting in more consistent growing conditions and perhaps better yields.

One of the primary benefits of indoor raising is the ability to maintain ideal water temps all year round. This is especially advantageous in areas with harsh weather conditions. Indoor systems may employ heaters, insulation, and temperature control to provide a stable environment for tilapia development.

Indoor systems make it simpler to regulate water quality since they may adopt sophisticated filtration and recirculating aquaculture systems (RAS). RAS technology reduces water use and enables continuous recycling and treatment of water, ensuring high-quality conditions.

Indoor systems also allow for year-round production, unaffected by seasonal variations. This may result in a regular supply of tilapia, allowing for better fulfillment of market needs.

However, indoor raising has greater initial and ongoing expenditures. Climate control and filtration need considerable amounts of infrastructure, equipment, and energy. In addition, space limits may limit production size as compared to outdoor systems.

Outdoor Tilapia Rearing

Outdoor rearing solutions, like as ponds or cages in natural water bodies, provide a cost-effective and natural setting for tilapia production. The initial expenditure is often minimal, and natural resources like as sunshine and water flow may be harnessed.

Outdoor rearing has the benefit of providing natural food sources like plankton and water plants. This may augment commercial feeds while lowering feed expenses. The increased area in outdoor systems also enables for greater stocking densities, which may increase productivity.

Outdoor systems benefit from natural processes that assist in preserving water quality. Sunlight supports photosynthesis, which increases oxygen levels, whilst natural water flow assists in waste elimination. However, regulating water quality might be more difficult owing to outside influences such as weather, runoff, and possible pollutants.

Temperature and water conditions may vary each season, affecting growth rates and output cycles. Tilapia in outdoor systems may develop at a slower rate during colder months or in bad weather conditions. Predation by birds or other animals may also be an issue in outdoor systems.

Finally, climate, available resources, production objectives, and budget all influence whether to raise animals inside or outside. Some producers may choose to combine the two systems to balance the advantages and offset the drawbacks of each.

In conclusion, successful tilapia farming requires good stress management, modern water filtration and aeration technologies, efficient pond management, and careful consideration of rearing settings. Farmers that apply these methods may improve tilapia health, growth, and production, resulting in a more sustainable and lucrative enterprise.

CHAPTER SEVEN

Ten Tilapia Growth Stages

1. Fry stage: Newly born tilapia that feed mostly on yolk sac reserves.

2. Fingerling Stage: Small fish begin to accept prepared diets.

3. Juvenile Stage: Rapid development needs adequate diet and water quality.

4. Sub-adult Stage: Nearing maturity, requiring close monitoring for gender distinction.

5. Adult Stage: Mature fish ready for breeding or marketing, depending on the farm's goals.

6. Spawning Stage: Fish acquire sexual maturity and begin breeding activities.

7. Egg Stage: Fertilized eggs cling to surfaces for incubation.

8. Larval Stage: Eggs hatch, and larvae ingest yolk sacs before eating.

9. Fry Stage (post-larval): Fish develop into free-swimming individuals and begin external eating.

10. Grow-out Stage: Growth continues until harvest size, which is maintained for optimum weight increase and health.

Steps to Follow When Harvesting Tilapia

1. Monitoring Growth: Evaluate fish size and readiness for harvesting.

2. Water Quality Check: Ensure that the conditions are appropriate to minimize stress while handling.

3. Harvest Tools: To harvest fish safely, use proper nets or traps.

4. Sorting by Size: Separate the fish to guarantee consistent marketability.

5. Humane Handling: Reduce stress during transportation and processing.

6. Sedation (optional): If required, calm the fish using gentle means.

7. Harvesting: Remove fish from ponds or tanks in an effective manner.

8. Cleaning: Rinse the fish to remove dirt and excess mucous.

9. Quality Control: Look for any evidence of sickness or deformity.

10. Chilling is optional. To keep it fresh, use ice or cooled water.

11. Packaging: Prepare for transportation or rapid processing.

12. Maintain the cold chain to retain quality.

13. Transport: Ensure prompt delivery to the market or processing facilities.

14. Processing: Fillet or package fish based on market need.

15. Quality assurance involves monitoring post-harvest quality and customer satisfaction.

Ways to Process and Store Tilapia

1. Filleting involves removing bones and preparing fillets for fresh eating.

2. Freezing: Use quick freezing to maintain freshness and improve shelf life.

3. Smoking is a traditional technique of taste improvement and preservation.

4. Drying: Dehydrating fish to make shelf-stable goods.

5. Canning is the process of sealing fish in cans or jars for long-term preservation.

6. Salting is preserving fish in salt for a longer shelf life.

7. Pickling is the process of preserving and flavoring foods using vinegar or brine.

8. Vacuum packing involves removing air to avoid oxidation and spoiling.

9. Fermentation is the process of using natural microorganisms to preserve food and generate flavors.

10. Cold storage involves keeping fish at low temperatures to slow down microbial development.

CHAPTER EIGHT

Sustainable Tilapia Farming Practices

1. Water Management: The effective use and conservation of water resources.

2. Aquaponics combines fish aquaculture with hydroponic plant cultivation.

3. Recirculating Aquaculture Systems (RAS) reduce water exchange and waste.

4. Organic Feed: Feed that has been responsibly obtained and certified organic.

5. Integrated Pest Management (IPM) uses natural approaches to manage pests and illnesses.

6. Habitat restoration entails improving natural environments for biodiversity.

7. Efficient Energy Use: Using renewable energy sources whenever feasible.

8. Waste Management: Recycling and reusing waste products on the farm.

9. Community engagement entails involving local communities in sustainable practices.

10. Certification: Obtaining certificates for sustainable aquaculture operations.

11. Genetic improvement involves selective breeding for disease resistance and growth.

12. Education and training: Farmers may learn best practices continuously.

13. Monitoring and control include regular assessments of water quality and fish health.

14. Diversification entails cultivating several species to improve environmental equilibrium.

15. Buffer zones protect water bodies from agriculture runoff and contaminants.

16. Carbon footprint reduction involves lowering greenhouse gas emissions.

17. Sustainable sourcing is the ethical procurement of resources such as feed and equipment.

18. Technology Adoption: Using IoT and AI to improve agricultural management.

19. Policy advocacy entails supporting policies that encourage sustainable aquaculture.

20. Transparency: Giving customers clear information regarding agricultural techniques.

Economic Considerations for Tilapia Farming

1. Market Demand: Evaluating customer preferences and trends.

2. Production costs include charges for feed, labor, and infrastructure.

3. Price Fluctuations: Monitoring market prices to ensure profitability.

4. Economies of scale include scaling activities to lower per-unit expenses.

5. Financial planning includes budgeting for seasonal fluctuations and unanticipated expenses.

6. Risk management includes mitigating hazards associated with weather, sickness, and market fluctuations.

7. Government Support: Obtaining subsidies or incentives for agricultural growth.

8. Insurance protects against losses caused by disease epidemics or natural calamities.

9. Marketing strategy refers to the successful promotion of items to target markets.

10. Value-Added items: Create premium items with better margins.

11. Labor Efficiency: Improving worker productivity and training.

12. Transportation Costs: Cutting costs in distribution logistics.

13. Storage Facilities: Investing in appropriate cold storage to ensure product quality.

14. Certification costs: Creating a budget for certifications to get access to premium markets.

15. Seasonal Variability: Preparing for changes in output and sales.

16. Credit and financing: Obtaining loans or credit lines for growth.

17. Market Channels: Comparing direct sales to wholesale marketplaces.

18. Legal compliance entails adhering to agricultural and commercial regulations.

19. Analysis of rivals and market positioning.

20. Customer Relationships: Developing loyalty via quality and service.

21. Technology investments include adopting efficient agricultural and processing technology.

22. Supply Chain Management: Ensuring that inputs arrive on time and outputs are efficient.

23. Long-term sustainability means balancing short-term advantages with long-term viability.

24. Labor Regulations: Complying with labor laws and fair employment practices.

25. Financial Reporting entails keeping correct records for financial analysis.

26. Market research is the process of gathering information about customer preferences and behavior.

27. Collaboration: Working with suppliers, distributors, or research organizations.

28. Brand Development entails creating a strong brand identity in the market.

29. Customer Feedback: Utilizing feedback to enhance product offerings.

30. Exit Strategy: preparing for future changes in market conditions or personal situations.

This thorough review covers all elements of Tilapia farming, including development phases, harvesting procedures, sustainable practices, processing processes, and economic concerns. Each part highlights the need for thorough planning, effective management, and adherence to sustainable principles for Tilapia farming success.

Marketing Your Tilapia: Examples

Marketing tilapia includes knowing customer preferences, developing a brand identity, and implementing successful advertising techniques. Digital marketing tools like social media platforms, websites, and online marketplaces may help you reach a wider audience. For example, tilapia farms may demonstrate sustainable methods to attract environmentally sensitive customers.

Partnering with local eateries or grocery shops might help increase awareness. Packaging innovations, like as eco-friendly materials or simple portioning, might entice health-conscious consumers. Using certifications such as organic or sustainably sourced may also help distinguish items in the market.

CHAPTER NINE

Legal and Regulatory Implications of Tilapia Farming

Tilapia farming is subject to several laws, including water quality requirements, land use licenses, and environmental impact studies. Compliance ensures long-term operations while avoiding fines. Understanding the local, state and national aquaculture legislation is critical. For example, acquiring licenses for water consumption and disposal, following worker health and safety standards, and completing labeling requirements are critical.

International commerce may be subject to extra rules such as export certificates or tariff considerations, which might limit market access.

Innovations in Tilapia Farming

Tilapia farming innovations include genetic advancements for disease resistance and growth rates, sophisticated water filtration systems to preserve water quality, and automated feeding systems to maximize feed efficiency.

Integrated aquaponics systems combine fish farming and plant culture, using fish waste as fertilizer. IoT (Internet of Things) apps allow remote monitoring of water parameters and fish behavior, which improves farm management. Biotechnology developments like probiotics and immunization methods help to avoid illness, resulting in increased yields and healthier livestock.

Integrating Tilapia into Other Aquaculture Systems

Integrating tilapia with different aquaculture systems improves resource efficiency and expands product choices. Combining tilapia with shrimp or prawn cultivation in the same pond optimizes space use and increases revenue streams.

To successfully use fertilizer waste, integrated multitrophic aquaculture (IMTA) systems grow seaweed or shellfish alongside tilapia. Crop rotation with rice paddies benefits from tilapia's ability to manage pests and recycle nutrients. Such integrations need careful species selection, water quality control, and ecological balance to optimize reciprocal benefits.

1. Water Quality Issues:

- Problem: Poor water quality causes stress and sickness.

- Monitor pH, ammonia, nitrite, and oxygen levels routinely. Ensure proper filtration and water circulation.

2. Temperature fluctuations:

- Sudden temperature fluctuations might cause stress.

- Solution: Use heaters or cooling systems to keep water temperature steady. Slowly adapt fish to new temperatures.

3. Oxygen deprivation:

• Problem: Insufficient oxygen levels cause fish suffocation.

• Solution: Improve aeration and surface agitation. Ensure enough oxygen exchange with correct water movement.

4. Overcrowding:

• Problem: High fish density causes stress, aggressiveness, and disease transmission.

• Solution: Follow stocking density standards. Provide ample room for each fish to prevent competition and stress.

5. Poor Feeding Practices:

• Problem: Over or underfeeding might lead to health risks.

- Feed Tilapia in suitable quantities dependent on their development stage and water temperature. Quickly remove any uneaten food.

6. Disease outbreaks:

- The problem is the spread of viral, bacterial, or parasitic illnesses among fish.

- Solution: Quarantine new fish. Maintain proper water quality and cleanliness. Diseases should be treated quickly with proper drugs.

7. PH imbalances:

- Problem: Fluctuating or high pH levels may harm fish health.

- Solution: Regularly monitor pH levels. To keep Tilapia's pH at 6.5-8.5, use buffers or gradually change the water chemistry.

8. Ammonium Poisoning:

- Problem: High amounts of ammonia from fish excrement or rotting materials.

- Solution: Ensure effective biological filtration. Make frequent water changes. Avoid overfeeding to avoid ammonia emissions.

9. Nitrite Toxicology:

- Problem: High nitrite levels hinder fish respiration.

- Solution: Monitor nitrite levels. Ensure that biological filtration is enough to convert nitrites into less harmful nitrates. Make any required water adjustments.

10. Nitrate Accumulation:

- Problem: High nitrate levels lead to stress and impaired development.

- To regulate nitrate levels, examine them regularly and use effective filtering methods. Plants may also absorb nitrates.

11. Algal blooms:

• Problem: Nutrient imbalance leads to excessive algae growth.

• Solution: Limit nutrient intake by overfeeding or organic materials. Use algae-eating fish or physically remove the algae. Consider ultraviolet sterilization.

12. Predation:

• Problem: Tilapia are preyed upon by other fish and animals.

• Solution: Use nets or screens to secure ponds and tanks. Use physical obstacles to repel predators. Predators should be monitored and removed swiftly.

13. Handling stress:

• Problem: Rough handling might cause harm or stress to Tilapia.

- Solution: Use gentle handling methods for transfers and treatments. Reduce exposure to air while providing enough support for fish.

14. Breeding Challenges:

- Problem: Low reproductive success and fry survival.

- Solution: Maintain optimal breeding circumstances (temperature and water quality). If there is hostility, separate the breeding partners. Provide proper fry-rearing facilities and food.

15. Environmental stressors:

- Problem: Tilapia may be affected by loud sounds, vibrations, or rapid changes in illumination.

- Solution: Reduce disruptions around rearing areas. Provide constant light and dark cycles. Covers or blinds may help to mitigate environmental swings.

16. Inadequate filtration:

- Insufficient filtering causes poor water quality.

- Solution: Invest in proper filtering systems (mechanical, biological, and chemical). Size filters based on tank or pond capacity and fish population.

17. Temperature shock:

- Problem: Rapid temperature fluctuations during water exchanges or transfers.

- Solution: Maintain consistent water temperature during transfers. To prevent shock, gradually adjust the water temperature in unfamiliar situations. To keep temperatures stable during changes, use heaters or chillers.

18. Chemical contaminants:

- Problem: Pollutants and chemicals in water may harm Tilapia's health.

- Regularly test water for pollutants. Use dechlorinators if required. Pesticides and chemicals should not be used near animal-raising sites. Obtain water from pure, uncontaminated sources.

19. Electrical Issues:

• Problem: Malfunctioning equipment and electrical risks near water.

• Solution: Regularly inspect and maintain electrical components. To avoid potential risks, use grounded outlets and waterproof equipment. Implement safety practices to quickly respond to electrical outages.

20. Malnutrition:

• Problem: Nutrient deficiencies may cause stunted development and health risks.

• Solution: Offer a balanced meal with enough protein and vitamins. Supplement with commercial feeds or naturally occurring items like as vegetables and tiny insects. Assess fish development and health regularly and alter the food as required.

21. Behavioral Issues:

• Problem: Tilapia exhibits aggressive, bullying, or aberrant behavior.

• Monitor fish behavior frequently. Provide appropriate hiding places or shelters to deter aggressiveness. Separate aggressive people as required. Provide enough space and resources to reduce rivalry.

22. Equipment Failure:

• Failure of pumps, aerators, or warmers harms fish health.

• Solution: Regularly maintain and replace equipment as required. Keep backup equipment on hand. Monitor performance regularly. Perform regular inspections to detect possible problems before they create failures.

23. Parasitic infestations:

• Parasites affecting Tilapia might be external or internal.

• Solution: Quarantine new fish. Use suitable treatments to treat infestations. Maintaining adequate water quality will lessen stress and vulnerability. Perform frequent health inspections to detect pests early.

24. Fungal infections:

• Problem: Fungal development in fish caused by poor water quality or injuries.

• Solution: Enhance water quality and cleanliness. Use antifungal drugs to treat infected fish. Isolate-affected people if required. Address the root causes, such as injuries or inadequate environmental conditions.

25. Genetic Issues:

• Inherited genetic diseases affect Tilapia health and reproduction.

• Solution: Select breeding stock wisely. Keep an eye out for any anomalies in offspring. Consult geneticists or professionals for help with breeding techniques. Use selective breeding to improve desired features while minimizing genetic concerns.

26. Stress from Handling Tools:

• Improper usage of tools may lead to injury and stress.

• Solution: Use proper tools (nets, containers) while handling fish. Avoid unexpected movements. To reduce stress on fish, provide sufficient support. Teach workers about gentle handling procedures.

27. Waterborne Toxins:

• Problem: Toxins or contaminants in water harm Tilapia.

• Solution: Regularly test water for contaminants. To eliminate poisons, use activated carbon or other absorbent materials. Ensure that water sources are pure and uncontaminated. To avoid the entry of toxins, implement strong biosecurity procedures.

28. Disease Resistance Issues:

- Problem: Tilapia has limited resistance to common illnesses.

- Solution: Choose disease-resistant strains for breeding. Maintain high-quality water and diet to improve immunological function. If immunization regimens for particular illnesses are available, implement them.

29. Waterborne pathogens:

- Problem: Potentially dangerous bacteria, viruses, or protozoa in water.

- Solution: Reduce pathogens using UV sterilizers or ozone systems. Place fresh fish in quarantine. To prevent spread, use biosecurity precautions. Monitor fish health and water quality regularly to recognize and manage disease concerns as they arise.

30. Monitoring and Record Keeping:

- Problem: Inconsistent monitoring and record-keeping might lead to unsolved concerns.

- Set a regular monitoring routine for water conditions, fish health, and equipment. Maintain careful records of observations, treatments, and results. Utilize data to discover patterns and enhance management procedures.

Addressing these frequent tilapia-rearing issues requires persistence, proactive management, and a detailed grasp of the aquatic environment and fish behavior. Aquaculturists may improve Tilapia health and assure long-term production by applying specialized treatments for each concern.

CHAPTER TEN

Future Trends in Tilapia Farming

1. Technological Advancements in Recirculating Aquaculture Systems (RAS) RAS technology is projected to gain popularity because of its water-saving efficiency and reduced environmental effects. Innovation in this area will result in more sustainable and scalable tilapia farming operations.

2. Genetic improvement programs will prioritize selective breeding and genetic engineering to improve growth rates, disease resistance, and feed conversion ratios. These advances will boost output while reducing the environmental impact of tilapia aquaculture.

3. Alternative Protein Sources for Feed The development of alternative feed components, such as insect meal and algae, would lessen reliance on conventional fishmeal and soybean meal, therefore boosting industrial sustainability.

4. Integrated Multi-Trophic Aquaculture (IMTA) systems, in which tilapia farming is combined with other species such as shellfish and seaweed, will become increasingly widespread. This strategy increases resource efficiency and environmental sustainability.

5. Blockchain technology will improve traceability and transparency across the tilapia supply chain. This will increase customer trust and enable better control of food safety and quality.

6. Climate Change Adaptation Strategies: As climate change affects water temperatures and availability, tilapia farms will use heat-tolerant strains and improve water management procedures to promote resilience.

7. Urban aquaculture, which uses city space to cultivate tilapia, will expand. This trend capitalizes on closeness to markets, lowering shipping costs and assuring fresher goods.

8. Enhanced Biosecurity Measures Stronger biosecurity standards will be introduced to avoid disease outbreaks, protect fish health, and maintain consistent output levels.

9. Consumer Demand for Sustainable Products Increasing consumer knowledge and demand for sustainably produced seafood will encourage farms to adopt more ecologically friendly techniques and get certifications demonstrating their commitment to sustainability.

10. Government and Policy Support: Improved government policies and support for sustainable aquaculture techniques, such as subsidies, research funding, and regulatory frameworks, would help the tilapia farming business develop and thrive.

Community and Social Advantages of Tilapia Rearing

Economic Development Tilapia cultivation creates job possibilities in rural and coastal locations, which helps to boost local economies. It produces employment not just in agriculture, but also in associated industries like feed production, processing, and marketing.

Food Security Tilapia is an important source of protein, critical nutrients, and low-cost food for many populations. Its cultivation may assist in reducing food insecurity by providing a consistent supply of nourishment.

Women's Empowerment In many locations, women participate in many phases of the tilapia-producing process. This inclusion strengthens women economically and socially, allowing them to play more active roles in community development.

Education and Skill Development Tilapia farming efforts often incorporate training programs to provide local farmers with vital aquaculture skills. This information transfer improves communities' ability to participate in sustainable and productive agricultural techniques.

Sustainable Livelihoods Tilapia cultivation may offer a sustainable livelihood choice for small-scale farmers, eliminating dependency on overfished wild populations while also encouraging environmental protection.

Community Cohesion Aquaculture projects often promote community cooperation and cohesion as people work together to achieve shared objectives. This concerted effort fosters social relationships and promotes community resilience.

Youth Engagement Tilapia farming may include young people in profitable and creative agricultural activities, hence lowering youth unemployment and rural-urban migration.

Health Benefits Increased availability of fresh, locally produced tilapia promotes dietary variety and improves health outcomes in communities.

Environmental Awareness Community members become more conscious of environmental protection and responsible resource management when they participate in sustainable tilapia farming.

Cultural Integration Tilapia farming may be combined with traditional agricultural methods, conserving cultural legacy while introducing modern aquaculture technology.

CHAPTER ELEVEN

Frequently Asked Questions About Tilapia Rearing

1. What are the optimal water conditions for tilapia farming? Tilapia grow in warm water with temperatures ranging from 25 to 30°C. The pH should be neutral to slightly alkaline (6.5-8.5), and the dissolved oxygen concentration should be more than 5 mg/L.

2. How long does it take for tilapia to reach market size? Tilapia typically reach market size (about 500 grams) in 6-8 months, depending on water quality, feed, and farming procedures.

3. What do tilapia eat? Tilapia are omnivorous and may consume a wide range of foods, including commercially designed pellets, plant debris, and tiny aquatic creatures. High-protein diets encourage rapid development.

4. How do I maintain water quality in tilapia farming? Regular monitoring of water characteristics including temperature, pH, dissolved oxygen, and ammonia levels is critical. Aeration, filtration, and frequent water exchanges all assist in keeping conditions ideal.

5. What are the most prevalent illnesses in tilapia, and how can you avoid them? Bacterial, parasitic, and fungal infections are among the most common illnesses. The majority of infections may be avoided by practicing good biosecurity, preserving water quality, and eating well.

6. Can tilapia be cultivated in both freshwater and saltwater? Tilapia are typically freshwater fish, however they can withstand brackish water with salt levels as high as 15 ppt. Some species, such as the Mozambique tilapia, can tolerate greater salinity.

7. How can I establish a small-scale tilapia farm? Begin with site selection, assuring the availability of clean water and a suitable climate. Set up tanks or ponds, get

fingerlings from a reliable hatchery, and devise a feeding and management strategy.

8. What is the stocking density of tilapia? Stocking density varies depending on the agricultural method. Extensive systems typically hold 2-4 fish per square meter, although intense systems may tolerate larger densities with sufficient aeration and filtration.

9. How can I increase the growth rate of tilapia? To improve growth rates, provide high-quality, nutritionally balanced feed, keep water conditions appropriate, and guarantee enough space and oxygen levels.

10. Are there any restrictions governing tilapia farming? Regulations differ by nation and area. Farmers must follow local aquaculture laws, and environmental restrictions, and get the required permissions to operate.

Conclusion

Tilapia farming provides several potentials for economic growth, food security, and community empowerment. Successful case studies from across the globe demonstrate the potential for creative and sustainable methods. Future developments in the sector include technology breakthroughs, genetic enhancement, and alternate feeding.

Community advantages, such as job generation, education, and health improvements, highlight the social worth of tilapia farming. As the business expands, prospective farmers and stakeholders will need to answer commonly asked questions and make use of existing tools. Tilapia farming has a bright future, thanks to sustainability, innovation, and community participation.

www.ingramcontent.com/pod-product-compliance
Lightning Source LLC
Chambersburg PA
CBHW070344230526
45471CB00006B/2429